Amazing Oceans

Written by P.A. Lin
Illustrated by V.A. Kitsco

Dedicated to all of you who love the ocean!

*Beautiful, blue, sparkling oceans, vast and strong,
Amazing oceans, filled with life, a history long.*

*Five major oceans surround Earth's precious land,
Five major oceans make beaches of rocks and sand.*

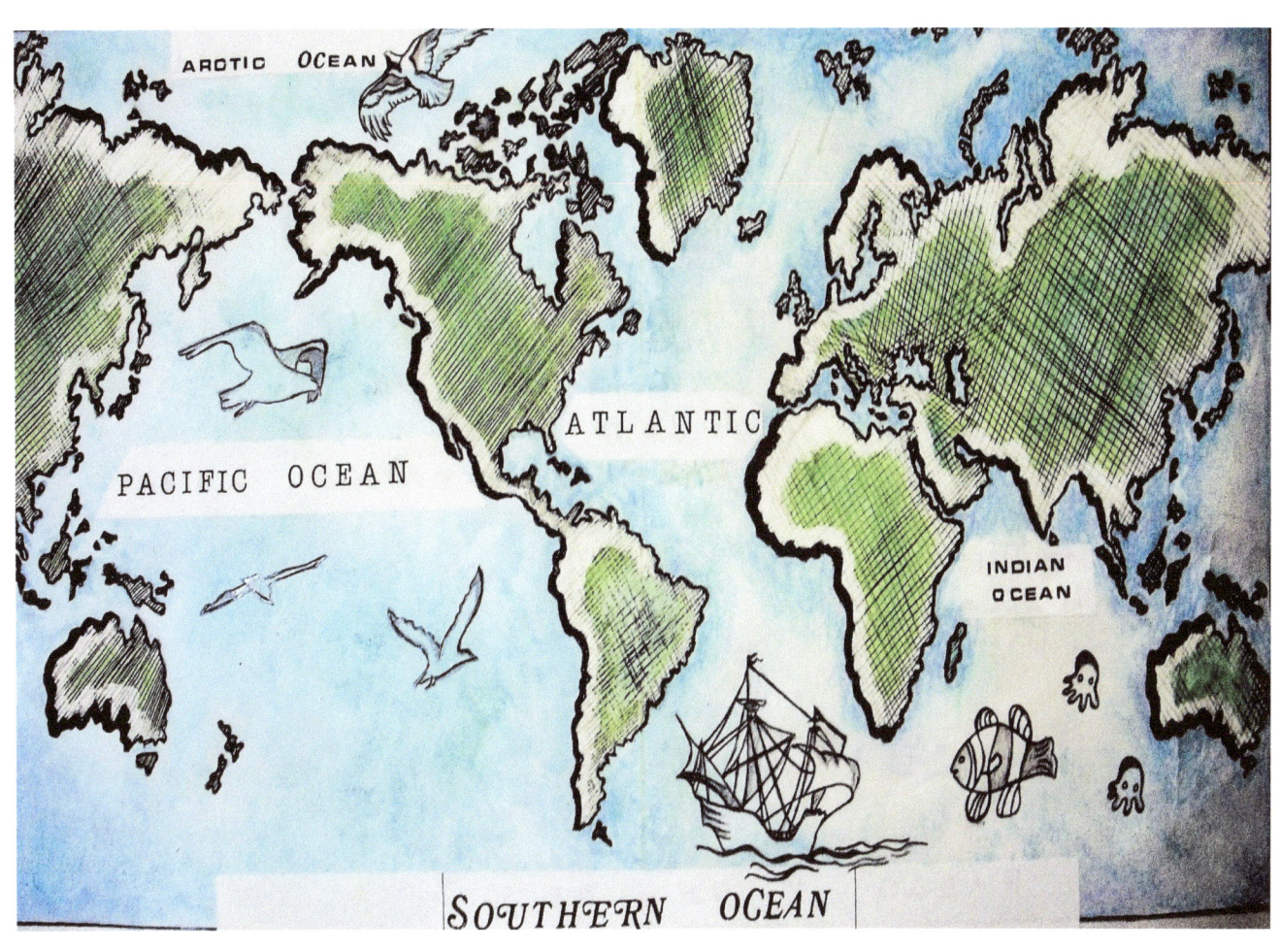

All of the oceans have tides, low and high,
Waves coming in to greet you,
then moving out to say goodbye.

*Large waves, crashing noisily, against the shore,
Splashing you with water as they hit the rocks and roar.*

It's breathtaking to stand, and look upon a wave.
It's also pretty grand to think of all the ocean gave:
Seaweed, seafood, fish, and such,
That's just to eat, it isn't much,
It's just a little part of the ocean's enormous heart.

The ocean's waters cover more than 70% of the Earth,
All are filled with saltwater, and all are of great worth.

It's vital that we work to keep the oceans clean,
If we dump stuff inside it, that's really very mean.

The ocean's not just pretty, it's many creatures' home,
Inside the ocean, an abundance of life forms roam.

Within every single ocean, sharks and whales make their home,
They need the five oceans that make up the marine biome.

The Pacific Ocean's famous for its many coral reefs,
The life-expectancy of which is most certainly not brief!
You may not realize that corals are very much alive,
They are animals, not plants, though they photosynthesize

Around the Great Barrier Reef of Australia, fish love to live, so free.
Home to many ocean creatures, even sea turtles you might see.

The Pacific Ocean, warm and blue, contains the ring of fire:
Where volcanoes erupt and earthquakes take place, it's really rather dire.
It's a little bit ironic, then, that it means the "peaceful sea,"
Although, to look upon it, one cannot deny its great beauty.

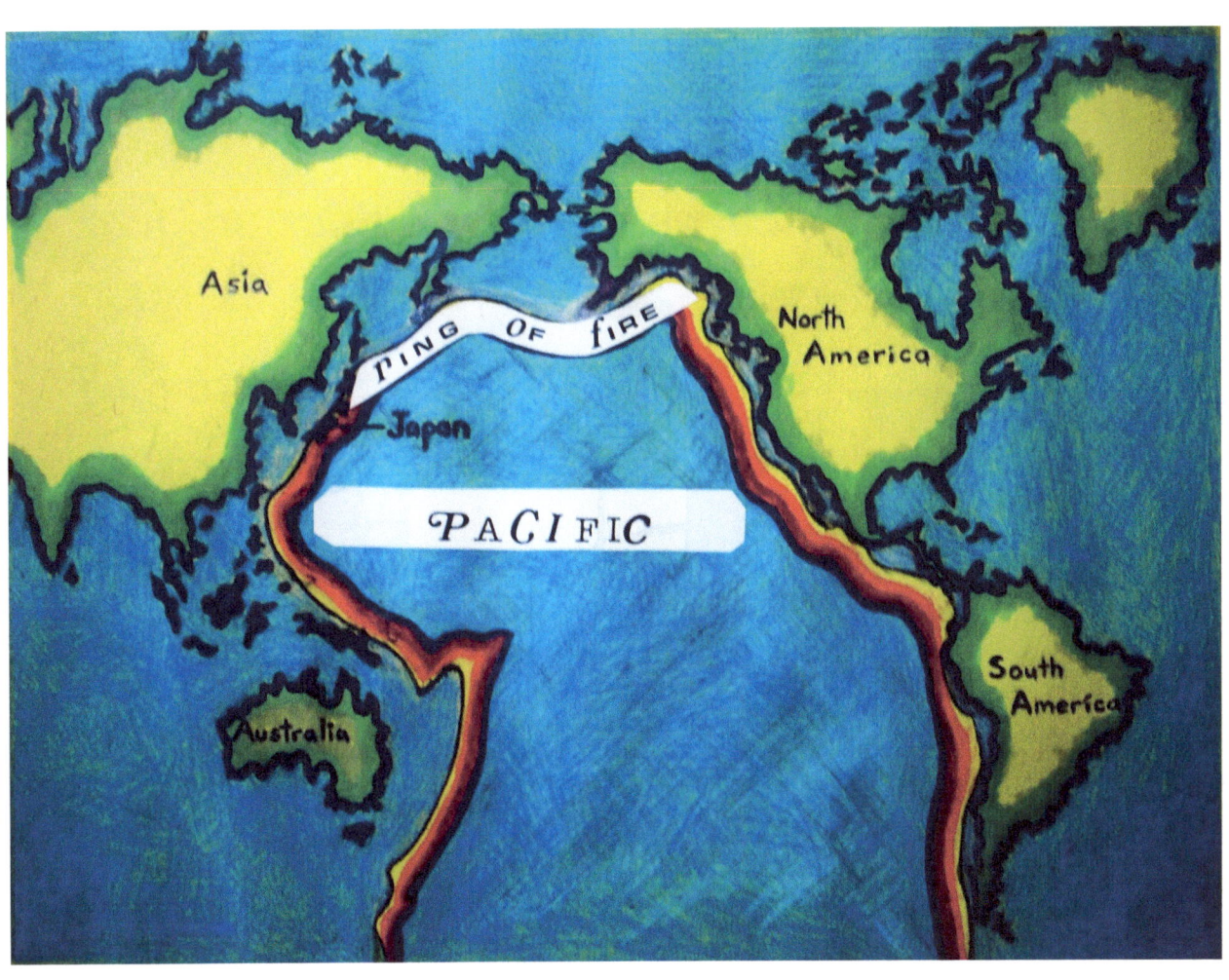

If you're lucky, you might see dolphins jumping high,
Dolphins jumping in the air, trying to reach the sky.
They like to live in the Pacific because it's pretty warm,
And with so much water to swim in, they seldom come to harm.

The second biggest ocean, the Atlantic, is also pretty great.
It's true that in these waters many sperm whales like to mate.

The special part of this ocean is the Mid-Atlantic Ridge,
The largest mountain range on earth, it stretches like a bridge,
It stretches all the way from Iceland, in the north, so cold,
All the way down south to Antarctica, beneath the Atlantic,
Oh, so bold!

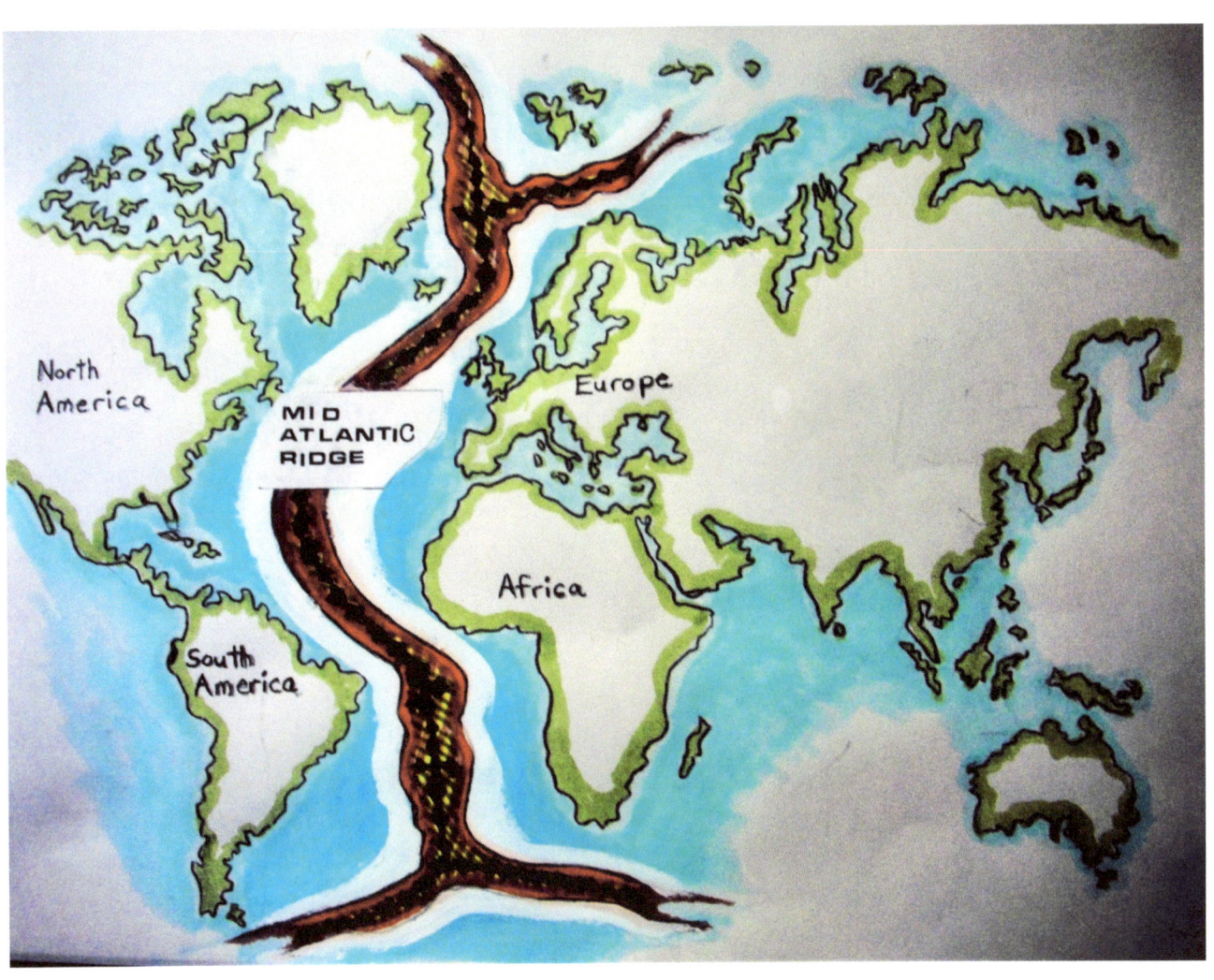

It means the "Sea of Atlas", a name that comes from Greek.
And it happens to be the ocean in which the Titanic sprung a leak!
Atlas was a Titan God forced to hold the heavens on his shoulders.
He must have been as strong as several giant boulders!

Africa, Asia, and Australia hug the Indian Ocean's shores,
The third largest ocean, it opened up the doors
Of trade between countries for tea and for spice,
Not to mention that humpback whales think it's really quite nice!

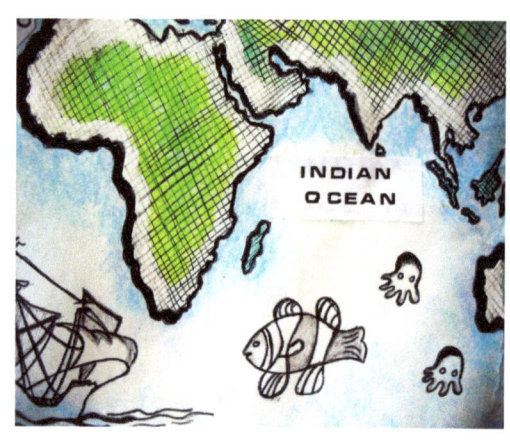

The Southern Ocean, though lovely, is cold and it's gray,
Bordered by Antarctica where Emperor Penguins do stay.
It's home to blue whales, which are really quite large,
And the plankton here bloom and really take charge.
Plankton are small living things that some animals eat,
Some can only be seen with a microscope, which is really quite neat!

The world's smallest ocean,
The Arctic Ocean,
comes from "Arktos," which means bear.
Four hundred fish species live in it
And of fishermen must beware.
Popular with whales, four species live here:
Bowheads, gray whales, narwhals, and belugas
Roam without fear.
It's up in the north,
Where the weather is cold,
And many big glaciers
Its surface does hold.
Polar bears play on icebergs that don't sink,
To them they are giant ice skating rinks.

There are many big problems all the oceans do face.
To fix them in time is a bit of a race.
Some marine creatures are endangered, you see,
Due to over-fishing and hunting, and lots of debris.

When leaks happen from big tankers shipping oil,
All the fish and marine life in the ocean recoil.
But most of the problems begin on the land,
Chemicals, sewage, and garbage
End up in the oceans' great hands.

The ocean's not meant to hold plastics and trash,
It makes the ocean dirty and with marine life does clash.
Garbage is bad for the life in the seas,
In a way it ends up harming both you and me.

So, next time you sit upon the ocean's sand,
Think about how you might lend a helping hand.

How much do you remember?

Short Answer Questions Based on this Book:

1. What are the names of the five major oceans on Earth? If possible, list them in order by size, from largest to smallest.
   a) _____
   b) _____
   c) _____
   d) _____
   e) _____

2. What percent of the Earth is covered by the ocean?

3. Are corals plants or animals? What makes it confusing to decide?

4. List three facts about the Pacific Ocean:
   a) _____
   b) _____
   c) _____

5. List three facts about the Atlantic Ocean:

   a) _____

   b) _____

   c) _____

6. List three facts about the Indian Ocean:

   a) _____

   b) _____

   c) _____

7. List three facts about the Southern Ocean:

   a) _____

   b) _____

   c) _____

8. List three facts about the Arctic Ocean:

   a) _____

   b) _____

   c) _____

Definitions

Biome: A biome is a big, naturally occurring community of flora and fauna (plants & animals) occupying a major habitat.

Coral: A large, stony substance secreted by certain marine invertebrates (creatures without a backbone) as an external skeleton.

Coral Reef: A ridge of rock in the ocean formed by the growth and deposit of coral.

Debris: Scattered pieces of waste or remains

Endangered: at serious risk of becoming extinct

Glacier: a slowly moving mass or river of ice created by the build up and compaction of snow on mountains or near the poles

Iceberg: a large floating mass of ice detached from a glacier and carried out to sea

Marine Biome: a naturally occurring community of aquatic plants and animals in the ocean

Definitions continued ....

Photosynthesize: when plants or certain living organisms use sunlight to make food from carbon dioxide and water.

Plankton: small, microscopic organisms that drift or float in the ocean or fresh water.

Tide: the alternate rising and falling of the ocean, usually twice in each lunar day, at a particular location, due to the attraction of the moon and sun.

Titanic: a luxury steamship that sank in 1912 after being hit by an iceberg on its first voyage.

Amazing Oceans Word Search

```
A B R E A T H T A K I N G B T C
N A R W H A L D E F G H I J I K
C R Q W H A L E S E P O N M T L
O S I C E B E R G E K E C I A P
R S A L T W A T E R C M I N N L
A T U V A W T X Y Z I O T D I A
L K J V I H I G F A F I C I C N
D L E M N O D P Q B I B R A D K
A S O U T H E R N O C E A N O T
E H X W V U S T R C A A E O L O
H O Y F E H G H I D P C D C P N
W R Z M A R I N E E J H C E H Y
O E A R D P O N M F L E B A I X
B C K B E L U G A G A S Z N N W
  P H O T O S Y N T H E S I Z E V
  U T S R A C I T C R A T N A Q P
  A T L A N T I C I H J K L M N O
```

Words:

ANTARCTICA
ARCTIC
ATLANTIC
BEACHES
BELUGAS
BIOME
BOWHEAD
BREATHTAKING
CORAL
DOLPHIN

ICEBERG
INDIANOCEAN
MARINE
NARWHAL
PACIFIC
PLANKTON
PHOTOSYNTHESIZE
REEF
SALTWATER
SHARK

SHORE
SOUTHERNOCEAN
TIDES
TITANIC
WAVES
WHALES

BONUS Coloring Page!

About the Author:

Patricia A. Lin (B.Sc., B.Ed., M.Ed.) is an educator in Calgary, Alberta, Canada, where she resides with her husband, daughter, and dog. When she is not working or writing, she loves spending time out in nature with her family. She is passionate about topics related to nature and the preservation of the Earth.

About the Illustrator:

Victoria A. Kitsco, a retired teacher with a B.A. and a professional diploma in Education, resides in Edmonton, Alberta, Canada. She has enjoyed art all of her life. She is also an avid gardener and a talented pianist. When she is not gardening or doing art, she loves to spend time with her family, including her granddaughter.

Please note:

All rights reserved, including the right of reproduction in any form, without permission of the author and/or illustrator.

The author can be contacted at plinauthor@gmail.com

The artist can be contacted at VickyArtist70@gmail.com

If you enjoyed this picture book, you may also enjoy these books, written by the same author and illustrated by the same artist:

Incredible Trees

A Mother's Love

Remember to consider posting a review on Amazon if you enjoyed the book, or feel free to contact the author and/or artist!

Namaste ☺

www.ingramcontent.com/pod-product-compliance
Lightning Source LLC
Chambersburg PA
CBHW040412220526
45473CB00004B/1210